FREE IS BAD

HOW THE FREE WEB HURT PRIVACY, TRUTH AND DEMOCRACY

AND WHAT YOU CAN DO ABOUT IT

JOHN MARSHALL

Free Is Bad TLDR - 1st Edition

Marshall, John Free Is Bad. - 1st ed. - 2020.

ISBN 978-0-578-81451-3

Orthogonal Thinking

For my parents, who gave me the ability to think, and the confidence to do so.

INTRODUCTION 1
1. SEARCH 7
2. EMAIL 17
3. AD TECH 27
4. SMARTPHONES 37
5. SOCIAL MEDIA 49
6. NEWS 59

INTRODUCTION

FREE IS BAD

It's a simple statement, that represents a somewhat more complicated concept.

I came to this conclusion after years of working in the ad-tech industry; founding startups, patenting technologies, and building tools to help businesses advertise more effectively.

After hanging up my spurs, I wrote a 300 page book about how that industry is now broken, not only hurting our privacy but hurting democracy.

It is, I think, a fascinating look at the long history of 'free' on the web. From search to email to social media and news, it explains why we think those things should be free, and why they can't actually be free. And what we can do to protect ourselves from the harm that reality causes.

I also live in the world, and I recognize that not everyone has the time or inclination to read about where email and search came from. What they want, simply, are the critical steps to understanding the threat and protecting themselves.

So I wrote this book too. The **TL;DR** (Too Long, Didn't Read) version of my original book, setting aside most of the history and research, to provide you with the core knowledge and step-by-step tools you need to protect yourself online.

It goes something like this:

Once upon a time, the web was free, designed for academic colleagues to trade knowledge and research.

Then one day the web became a marketplace. And like

any marketplace, it traffics in three things: *Sellers*, *Buyers*, and the *Products* they trade money for.

The free things we love, like email, search engines, social media and news, actually cost money to provide.

If we're not paying, who is? The ad-tech industry. Ad-tech funds our 'free' online services by buying you and me: our attention, our data, and our behavior. And they manipulate what we see and how we act to maximize the clicks they get and the dollars they earn... even if it means hurting us and society.

It doesn't have to be that way.

Lots of web services, including news publishing — especially news publishing — would be more valuable, reliable, and honest (and less harmful) if we paid for them. And in many cases, we can.

We just don't.

But we have the power to become the customer and reclaim our agency in the marketplace of the web. Not just for email and search and social media. But in the marketplace of ideas, where trustworthy, evidence-based news arms us to be informed and powerful citizens.

Stop being the product, and start being the customer.

Because Free Is Bad.

Part I:
Online Services

"*Information wants to be free*" said the tech giants

1. SEARCH

Vacations
Top Vacations
Vacation info
Vacay Blog
Vacay Budget

Google didn't invent search, and they didn't invent the advertising business model that drives it. Its founders even started out opposed to the very idea of ads, saying:

"It could be argued from the consumer point of view that the better the search engine is, the fewer advertisements will be needed."

Back then, the established giant was Yahoo, who defined the 'free is good' business model. They attracted a huge audience by curating websites that had all the stuff we wanted to read — news, sports, message boards, etc. — and offering access for free inside a huge site called a portal.

Yahoo made good money by displaying graphical banner ads on pages in the portal. The more people that saw the ad, the more Yahoo could charge. Their goal was to keep people inside, looking at the ads they controlled. So, they didn't make searching the web a priority, because if people left the portal, Yahoo lost that ad money.

Google, on the other hand, was focused only on search. They thought portals were complicated and required too many people to create web pages and sell banner ads.

That left Google with a problem. How could they make money if the purpose of their free service was to direct people away from their site to another business?

The investors in Google were patient, figuring the smart guys would eventually figure out how.

Yahoo wanted to: Keep us inside a world of news, entertainment, email and search all under one roof they controlled.

Google wanted to: Give us an easy way to find things to watch, read, or buy, no matter where it existed online.

The challenge they both had was: Finding the best way to make money at it.

The web grew fast. Sites were easy to build, and everyone embraced "information wants to be free." We lost interest in portals, and wanted to search for content ourselves.

Yahoo saw that search was becoming important, so they bought some search companies. They even licensed Google's up-and-coming search technology, too.

Both companies recognized something special: when we search, we tell the search engine what we want. What if they could somehow charge advertisers when we clicked on a search result that matched our needs?

A small company called GoTo.com had already found a way to do just that: match people's searches to businesses and charge for the click that takes them to the website. This became known as Pay-Per-Click, or PPC.

Google saw the brilliance and copied GoTo's concept.

Jim Brock, Yahoo's first attorney, said, "... we understood that PPC was going to eat the internet [and become] the dominant and most effective advertising mechanism."

So, Yahoo bought GoTo.com and its patents, forcing Google to pay for using them. Armed with the right to use Yahoo's technology, Google continued to grow and quickly eclipsed its foe.

We got used to seeing ads while looking for information. After all, that's how the Yellow Pages had worked before the internet. Everyone got a listing in the phone directory, but those who paid got better placement and fancy ads. Why not the same with search?

Google and Yahoo knew they could make money by matching searches to businesses.

Pay-Per-Click became the way businesses wanted to pay. Google sold it like hotcakes.

And we got used to seeing ads.

Today, over 90% of searches worldwide are done with Google. Yahoo's a distant memory.

Google makes their money by selling the 'onward click' that directs interested people to advertisers. The advertisers love it: now their ads are bringing them traffic that's actively seeking what they're selling.

Of course, the more accurately Google can match a person's interests and intentions to the advertiser's offer, the more valuable the click becomes. That means the more Google knows about us, the better.

So, they started keeping track of all our search queries, from today, and last week, and last year, building a complete history of us and our online behavior. They track not just the ads we see and click, but also plain links on the web, even in our emails. And they use that profile to match us to more ads. Expanding the profile became so important that Google started making phones, so they could get data from those too.

The early internet pioneers showed that using search to match businesses to consumers, and having the business pay, is OK when it's transparent. Just like the phone book. It's not OK when the phone book snoops on our lives.

You're not forced to use Google, and there's a dozen alternatives that work great. Google just assumes you don't know about them, and that you haven't read this book.

So, let's begin by choosing a better search engine. One that's fair, promises no snooping and has simple ads.

Google collects data about you everywhere, from its own services like Android, Gmail and Maps, but also from independent websites.

Searching the web is driven by ads but doesn't have to be driven by snooping.

Fix Your Searches

My choice of search engine is DuckDuckGo. Yes, it has ads. But we've seen that paying for search any other way doesn't seem to work. DuckDuckGo promises no snooping. They don't peek at your emails, nor your maps, nor inside your smartphone. They make money in a simple and fair way: by matching your search query to an ad inside the search results page. Just once. Without keeping track of your search history. That's it.

Make the switch

You can change the search engine inside the browser settings. It's easy, though obviously Google prefers you don't know how. Just go to the settings screen for the browser software. Or ...

Here's a secret: Install an app called "DuckDuckGo Privacy Browser." It keeps searches private and cuts down trackers and other bad stuff. You'll find it in the Apple App store or Google Play Store.

Get Support

Finally, if you're worried about making a mistake, the creators of DuckDuckGo have provided a web page with links to their software. It's at **spreadprivacy.com**.

Use DuckDuckGo for searches free from snooping.

Install DuckDuckGo Privacy Browser for even tighter privacy and less shenanigans.

2. EMAIL

We still love email, even with all its faults. What we don't love is an algorithm that reads our personal messages. That's just creepy. The postal service doesn't do that, and neither does the phone company. So why does Google?

There's a couple of quirks in the history of email you should know, one of which became Google's biggest advantage.

When email was invented in the 70s, nobody thought to prevent snooping. The nerds just assumed nobody would want to. Messages were sent and stored as plain text. Kinda like sending snail mail, but if nobody had invented envelopes.

The second quirk is that it always came bundled as part of another service, not something we paid for separately. When we first got an internet connection, maybe through AOL, they bundled a free email account as an incentive.

And we got used to the idea it was free.

Then Yahoo offered free email as a way to keep us inside their portal, knowing we would come every few hours to look for new messages. They figured we might click over to other services, like weather or news, where the ads were. But they didn't look inside our email messages, because it felt invasive. Plus, it was technically difficult.

Google launched Gmail for free, gambling that we'd accept having an algorithm read our emails and show ads in exchange. And we did, because Gmail was so much better than the other email tools.

That gave Google a HUGE advantage.

When email was invented in the 70s, nobody thought to make sure it was private. The nerds just assumed it would be.

Google gained a huge benefit by assuming we would tolerate the idea it was reading our messages.

Remember we said that Google builds a history of our searches over time? It turns out this is only possible if we log in to the search engine. But who would want to do that? So, Google needed a way get us to log in, and stay logged in. The hook was ... Gmail.

That's right. Being logged in to Gmail is not just a way they mine data from inside the emails. When you're logged in to Gmail, you're really logged in to everything Google. It also allows Google to track us all day, every day. When we visit other websites, Google tracks us there. When we use Chrome, we're logged in there too. Because we use email so much, and don't want to retype our password every hour, we stay logged in and keep sending more and more data to Google.

These days Google doesn't mine the contents of our emails. It doesn't need to, because its web trackers see us visiting the websites that email us anyway.

Gaining control of your privacy means switching from Gmail. It's important to switch to a provider that promises it won't snoop. Remember that those 70s nerds made it possible to snoop, and any free email provider will likely do so. We shouldn't blame them - they need to make money somehow. If we expect our emails to be private, we need to be willing to pay something. Email companies have bills to pay, and it's best if we contribute. That's fair.

Switching to private email is not hard. Here's my step-by-step guide.

If you use Gmail then you're logged in to Google all the time. They collect vast quantities of data on you, across the web.

There's plenty of private and secure email companies, but be careful of free. They have to make money somehow.

Get a private email address

In most cases you can't keep your free email address and escape data collection. Google won't let you pay and become the customer. They'd rather keep you as the product and sell you over and over again.

Instead, you have to switch to a new email provider that's not driven by ads. The good news is there's plenty to choose from. Every day more and more people like you are switching to better email services and regaining their privacy.

You'll find they're easy to use, inexpensive, and you'll like having a more distinctive address. Here's my favorites:

Fastmail, **Gandi**, **HEY**, **Hushmail**, **Mailfence**, **OnMail**, **Posteo**, **ProtonMail**, **Runbox**, **Tutanota**, **Zoho Mail**

If I had to pick one, it might be **OnMail** or **HEY**. Compare the prices & features, signup and choose your new email address. Paying money is how you get privacy, because Free Is Bad.

Install any apps they provide. They work better than the generic mail client that came with your device.

Get familiar with your new private email service and give the new address to your friends and family. Then you can switch over the important online services you use regularly.

Choose a new email address from a service that guarantees privacy. They have to make money somehow, so expect to pay a little.

Install any apps the new service provides. Start using your new address and tell friends and family.

Update online services

Changing the email address for bank accounts and credit cards should be your top priority. You might have to add your new email address before removing the old one.

You'll want to tell your friends and family about the new address, but they're often busy and just use the old one out of habit.

So, here's a time saving tip: In your old Gmail account you can set a vacation responder. Usually this tells people who email you that you're at the beach or whatever. Instead, set the message to tell them about your new email address.

Finally, I expect you have a LOT of other online services tied to the old email address. And you might have lost track of who has that address on file. Luckily, these days most devices and browsers have a password manager to store and retrieve the logins for all that stuff. That's a great place to get a list of all those services you use but forgot about. Login and change the email address.

Should you keep Gmail around for junk stuff? I don't recommend it. If you do, you'll be logged in to Google and all those trackers and ads will follow you.

Update bank accounts and credit cards first. Then tell friends. Set a vacation responder on your old account to remind people about the new address. Your password manager has a list of all the other online services.

Don't keep the old Gmail account. Staying logged in exposes you to all those trackers.

3. AD TECH

Nasty ads chase us around the web, interfering with our reading and watching and apparently reading our minds.

Why did anyone think this was a good idea?

Nobody planned for it to be like this. It happened because of some accidental loopholes in the design of the web, and how businesses exploited them. Let's start at the beginning:

Tim Berners Lee (or Sir Tim, as the British Queen later dubbed him) invented the web in 1992. It took off quickly because it was simple, and since he was an academic, Sir Tim encouraged anyone to use his work. All he really wanted was for "information to be free."

Behind the scenes, Sir Tim was actually building on the work of an earlier project: Xanadu. Created by Ted Nelson in the 60s, Xanadu had the concepts of hypertext and hyperlinks between documents. It also had a clever system of copyright control and built-in payment of tiny amounts of money from readers to authors. Nelson believed his system must reward authors who take the time to write the words, and micropayments was the fairest way.

But for his web, Sir Tim ignored the idea of paying authors because in academia it's not a big deal. Sadly, as the web exploded across the world, the technology never caught up to include payment. The best we got was poorly integrated use of another technology: credit cards. That's why every time you buy something online it's funky and prone to errors and fraud.

Before the web, it was Ted Nelson in the 60s who invented hypertext and hyperlinks in Project Xanadu. It included a payment system where **readers paid authors for their work**.

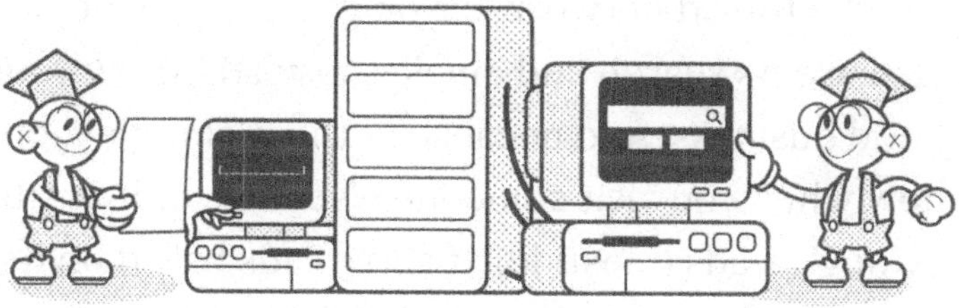

But the web was invented by academics, who saw no need to solve the payment problem. "**Information wants to be free**" they said.

The web won, and Xanadu is forgotten.

Sir Tim also decided that images on the web would be loaded after the text, to help with slow connections. When someone created a page, they left empty rectangles for images to come later. Sir Tim's design permitted those images to come from any web server, not just from the publisher's server. It didn't seem to matter to academics and nerds; they were more interested in the words, anyway.

The images loophole empowered ad networks to quickly become the source of money that still powers the internet today. They realized they could inject ads into those empty rectangles. And they made sure they would own the data collected via cookies, because that data also goes to their ad servers, not the website publisher. These third-party cookies and tracking scripts follow us everywhere, and it's big, big money. As the business press says, "data is the new oil."

Today the top ad network is, you guessed it, Google. If you also use its search and email, those ads know your history and business. And because of the rectangle loophole, they know when you visit someone else's site, if that site carries Google's ads (most do). It serves ads based on its data about you, and feeds data back into its algorithms. The data collection and flow of money is unbelievable.

For most websites, the only source of revenue is Google and other ad networks. Because Google pays the bills, they work hard to keep Google happy. They don't care about you beyond getting you to click an ad or another page with ads.

Ad networks used a loophole in the design of the web: images and cookies can come from anywhere, not just the site you're looking at.

If you use Gmail or Google Search, **their trackers follow you to other sites** and collect more data.

Once web publishers became dependent on ads, they started creating pages that would generate more money. They did this by:

- Putting more ads on the page
- Making some ads look like editorials and stories
- Creating emotive and divisive opinions

These tricks help the publisher generate clicks, which is how Google pays them. We shouldn't be surprised publishers do this. We may not like it, but we're not the customer.

Entrepreneurs saw how much money could be made from ad networks like Google and have rushed in to take advantage. After all, making a website is cheap, and it's easy to invent divisive opinions that get people engaged. In fact, lots of them steal the words someone else wrote. They just leave some empty rectangles in the pages, and networks like Google fill them with ads, and pay them.

Even if a site can't make enough money from real visitors clicking, they can create their own fraud clicks, using software robots.

It's become a huge problem. We don't like the nasty ads, trackers and shenanigans. Some people think legislation will fix it, but that hasn't happened so far. Europe and California are trying but it's not easy. Remember: these are loopholes in the fabric of the web. If we want to fix it, we're going to have to do it ourselves.

So, let's get started.

Some publishers create **divisive and emotional stories** because the ads are more profitable. Enraged people click more.

Fraud is also a problem, with sketchy publishers sending robots to click the ads. Google still pays them for those clicks.

Fix Your Browsing

Third party cookies is the name given to the biggest of Sir Tim's loopholes. Blocking them is tricky, and Google Chrome doesn't do it well. After all, it was designed to help Google track you. Here's my advice for better browsers:

Apple iOS: Safari is my first choice. It has good privacy settings and a secure password manager. And it comes with the device.

Google Android: I suggest you install DuckDuckGo Privacy Browser, or Brave. If you also have a Windows machine, consider Microsoft Edge.

Apple MacOS: Again my choice is Safari because it syncs everything with your mobile device.

Microsoft Windows: Consider Microsoft Edge. If you must use Chrome, install DuckDuckGo Privacy Essentials.

Block Ads?

If you block ads, you choke off the money that many web publishers need to survive. The checks from Google will stop. But still, too many ads are annoying.

A good compromise is Adblock Plus from eyeo. This extension blocks nasty, invasive ads but allows what eyeo calls Acceptable Ads that follow reasonable standards. And eyeo make a small amount of money from them, which helps support development. That seems fair.

Block third party cookies and trackers by dumping Chrome. Use Safari, Edge, or DuckDuckGo Privacy Browser.

Blocking all ads hurts the sites you visit. How are they going to survive? Instead use **Adblock Plus from eyeo**, which blocks nasty ads and permits nice ads.

4. SMARTPHONES

Smartphones started out as dumb phones. They were ugly and utilitarian pieces of plastic with funky buttons, teeny screens and they barely worked as phones. We loved the go-anywhere convenience but hated everything else.

We used them because we had no choice. The wireless carriers (Verizon, ATT, T-Mobile, etc.) controlled what devices we could buy and use. They told the phone makers what to build and blocked new features to keep them cheap. The carriers seemed to give us lots of options, but behind the scenes they were really stunting the technology and limiting us to what was best for them to lock us into expensive service contracts.

This ended in 2007 when Apple introduced the iPhone. It was so well designed and so much better than anything else, we wanted it no matter how hard the carriers tried to sell us something else. Apple built on that early success with a series of smart moves: apps, software updates, new models, and security features. The carriers would never have permitted this if Apple hadn't forced them to.

Google had bought Android the previous year and was planning a phone that would fit in with what carriers were selling: funky buttons and weird software. But they saw how we liked the smooth slab of glass and touchscreen so they changed course and copied that. Google could see that phones would be perfect for keeping us plugged into their world of search, email and advertising. Plus, those phones would feed even more data into their money machine.

Before smartphones, **carriers sold dumb phones**. They barely worked, but we had no proper choice. We got what they decided.

Apple changed this with the iPhone. It just worked better. **Google bought Android** and planned a way to copy Apple's success.

Google needed a way to catch up fast, so they made an incredibly smart move: give away Android to all those phone makers who were getting crushed by Apple. Remember, Google is an advertising company, so doing this was good business. They still made money when people used the phone for searching, email, and browsing websites. To make certain they'd capture our data, they forced makers to ship the phones preset to use Google services.

The makers like Samsung and Huawei who jumped aboard Android did okay, selling tons of phones but with razor thin margins. Those that didn't use Android, like BlackBerry and Palm, got left in the dust. Even Microsoft couldn't compete against Apple and Google, so they quit.

Choosing a smartphone is quite personal, and I know some people are Apple fans and stand in line for hours to buy the latest gadget. Other people think Apple is some weird cult, and sensible folk choose Android devices because they are nonproprietary. I'm not trying to pick sides, just explain a little of the history so you can make the best choices with the phone in your pocket.

Today all phones are basically a slab of glass with a grid of icons, and iOS and Android are your only choices. The biggest differences between them are driven by the fact that Apple and Google have different ways of making money.

As you know, "information wants to be free." At the same time someone needs to be the customer. The customer is either you or ad-tech.

Google just gave away Android, helping phone makers catch up with Apple.

But they forced the makers to ship their phones **preset to use Google** for search, email, etc. They knew the future ad sales would be huge.

Fixing your phone: iOS / Apple

The phone Apple sells is both secure and private. It comes with cloud services for photos, backups and other stuff, and those are private too. Apple won't let carriers or other companies install crappy software that you didn't ask for. It's a premium price phone because Apple treats you well.

Apple examines all apps that appear in the App Store to make sure they don't do really bad things, but you still need to make some choices about apps you install.

Where possible, use Apple. Delete Google Maps and use Apple Maps. It's private, and works great (but if you travel internationally, keep Google Maps in case). Use iCloud for photos, contacts etc. Delete Chrome and use Safari or DuckDuckGo (see below) for browsing.

Ditch and delete Gmail. I explained in the email chapter how to switch from Gmail. If you keep it on your phone, Google keeps you logged in to all their other invasive stuff.

Switch to DuckDuckGo. The easiest way is to install the DuckDuckGo Privacy Browser. Your searches go to DuckDuckGo instead of Google. Better yet, it blocks nasty trackers and keeps your data out of Google's clutches.

Trade in your old phone. Once a phone gets really old, maybe more than 5 years, Apple can stop issuing updates. You'll no longer get cool new features, nor bug fixes. Those fixes are actually a big deal, and stop you falling victim to scams. So, don't risk it. Just upgrade to a newer phone.

Apple devices are often more expensive. That's OK, because it means **you're the customer** and they don't sell you to ad-tech.

They inspect Apps to keep out the really bad stuff, but **you still need to make good choices**.

Fixing your phone: Android

All Android phones are tied to Google's advertising systems, and it's hard to escape. You'll be logged into Google's system and seeing ads... so they'll make money behind the scenes no matter what.

And it's not just Google and their own Pixel phones; there's hundreds of other makers shipping Androids. They can fiddle with the software, and bolt on their own even more invasive ad-tech shenanigans.

Or bad actors can get added later through apps you install. Google doesn't do a good job policing the Play Store.

I can't say "this brand of phone is good but watch out for these guys." Instead, let's divide the Android world into three tiers, and look at how they make money and how that affects us.

Tier 1: The phone is made by Google or a Google One partner. These are pure Android phones, and Google makes money from the ads. Of course, other Apps you've installed might be up to no good, depending whether they're free.

Tier 2: The phone came from someone you've heard of, like Samsung or Huawei. Google gets its ads into the mix, but the maker can add stuff too. Perhaps the NFL paid to stick their app on the home screen and share your data. Or the voice-assistant has a deal to recommend Tide detergent.

Tier 3: Cheapo guys you've not heard of. They can sell your data to anyone. They gotta make a buck, right?

Google doesn't **police the Play Store well**, and lets bad guys in. You need to be careful.

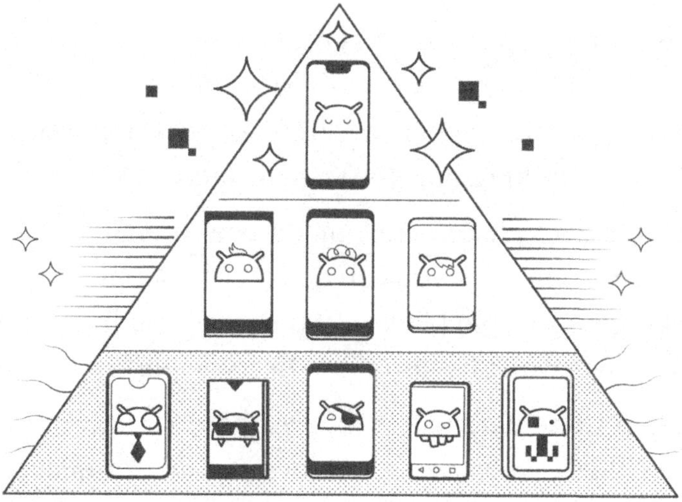

The likelihood of your phone coming with bad stuff **depends on how much it cost**.

Fixing Android cont'd

If your phone was suspiciously cheap, it's in tier 3 and you should trade up. Tier 2 at a minimum, ideally Tier 1.

You can't stop Android devices from sending data to Google, but you can reduce it. Here's how:

If you ditched Gmail in the earlier chapter, you'll be using a mail app supplied by the new, private service. That's a good start.

Chrome browser sends a lot of data to Google, so use something else like DuckDuckGo Privacy Browser. If your main computer runs Windows, use Edge from Microsoft on your phone.

Check every app. Android OS and the Play Store can't vet apps well, and those apps can do bad stuff. Delete things you don't recognize. Be especially careful of anything free. They need to make money, and you're their meal ticket.

Maps is tricky. Nerdy folk might use OpenStreetMap but it's not great. The best you can do is open Google Maps and select the option "Use Maps without an account." At least your searches won't be certain to become part of their data.

Keep it up to date with the latest patches and OS updates. It makes it less likely a scam catches you. Better quality phones receive updates for years. Cheapo phones might never get them. So, switch to a maker that takes updates seriously.

The big decision with phones is **Apple or Google**. There's a few weirdo phones but ignore them for now.

Apple is as secure and private as it gets.

Android is more complicated, and full privacy not possible. So, take steps to avoid the worst. **Buy Android direct from Google**.

5. SOCIAL MEDIA

We love social media for the connections but hate the negative effects it brings. Perhaps this surprises you, but I'm not going to tell you to delete your social media apps. Yes, it's only free because of the ads, and we know they build a profile of us using the data they gather, and use it to show us more ads. I don't love that stuff and I'm sure you don't either. But honestly, I like Facebook because I can easily stay in touch with friends and family. I treat it as entertainment, I limit the data collection where I can, and focus on the funny stuff.

I don't use Facebook for important things like news or stock markets. It's just not the right place to know the price of bitcoin, nor the results of an election. Funny videos, yes. Fiscal policy decisions, no.

I've weaned myself off most free services, like email, and I use private alternatives. I truly feel that Free Is Bad. But the 'social' part of social media means you need to be with other people, and right now everyone's on Facebook or something else with ads: Twitter, Instagram. So let's figure out how to get the good interactions with friends and family, and not get bogged down with conspiracy theories and negativity. Let's treat it as entertainment, like an episode of Seinfeld, and not as a source of facts for making important decisions.

From here on I'm going to say 'Facebook' to keep things simple, but it's the same on all social media platforms that are driven by ads.

It's true that **Facebook collects a ton of data** about you, and sells ads based on that. It's why the service is free.

We use it because that's where everyone else hangs out. Just **don't invite the bad stuff** along for the ride.

Get the good, without the bad

Don't doomscroll through bad stuff posted by negative people. Social media is just entertainment, and just like a bad movie, you can turn it off. Go ahead and delete the app if it's getting you down. (You can reinstall it later. Your contacts are stored in the cloud, so they'll reappear.)

Avoid politics and news on social media. We'll explain why in the next chapter, so for the moment just trust me. Skip past it, don't comment, don't react to it.

Next, prune your friends list. Get rid of people who behave badly, post political stuff, or are generally unpleasant. You really don't have to spend your precious time reading what they have to say. You can silence them for a while, or permanently, or unfriend them.

The more you react to political stuff, the more the algorithm will dish it up. It knows what you click.

Be thoughtful about how you react to posts. If other people will enjoy it, go ahead and share it. If people are going to have an angry reaction (and use the 'anger' or 'sad' emoticons) then don't post it. And don't use those negative reactions on other posts.

Think before you share, especially if your friends group includes people prone to conspiracy theories. Again, we'll cover fake news in the next chapter, and how it relies on being spread through social media.

Just don't make things worse.

Stay for the cat videos, **drop the toxic stuff**.

Prune your friends list to those you like.

Think before you share. Don't spread junk.

Close the loopholes

You know that Facebook collects data about you and uses that to serve ads. That's not a terrible trade-off. All that fun stuff for free. If you watch a video of a ranting parrot, who cares if you see ads for bird seed.

But suppose Facebook could see your activity on other websites? Suppose it knows if you visit a website with pictures of, er ... *unusual parrots*. Are you okay with Facebook knowing that? I think this is where most of us get uncomfortable. Not because of parrots of course, but the idea that Facebook sees all your web activity. Movies like the Social Dilemma have made the danger clear.

So how does Facebook know stuff that happens outside the app? Well, it's due to that little ad-tech loophole we talked about in chapter three: images and cookies can come from anywhere, or as we call them 'third parties'.

When you visit a website it often has little Facebook 'like' buttons. It's a handy shortcut so you can show support for the business. But it's also a sneaky way Facebook knows you're seeing the page. Those buttons are 'third party' so they share data without the site owner ever knowing, even if you don't click it. There is also totally invisible versions.

Fixing it is easy: a good privacy focused browser, as we explained in the previous chapter. Safari on Apple devices, Edge for Windows or DuckDuckGo Privacy Browser on Android.

Those 'like' buttons on the web leak data to Facebook unless you use a privacy focused browser to plug the loopholes.

Use social media for fun and connections, and **avoid the crazy people.**

Part II
The Media

"On the one hand information wants to be expensive, because it's so valuable. The right information in the right place just changes your life. On the other hand, information wants to be free, because the cost of getting it out is getting lower and lower all the time."

—Stewart Brand to Steve Wozniak, 1984

TL;DR

Information wants to be free.

Information also wants to be expensive.

That tension will not go away.

—Stewart Brand, 1987

6. NEWS

News is everywhere we look, all day and night. It's gone from a quiet and reliable place we could turn to for thoughtful information to a total crapfest of competing emotional opinions, shouting to be heard.

As we've seen with everything else online, news is just a business. If we understand how they make money, how that affects the quality of the news itself and then how that affects our lives, we can make smarter choices.

Of course, news existed before the internet, so let's start by taking a look at how they used to make money.

Newspapers were top dog in the 1800s, when printing became cheap. They became the de facto source for important information, and established the look we still recognize today: Banner headlines, editorials, advertisements and subscriptions. As the world changed through the Civil War, urbanization, and the rise of industry, newspapers kept people up-to-date, popularized new products via ads, and even helped them learn to read.

Newspapers had an unhealthy relationship with politicians until the 1900s when they found partisan bias was limiting the size of their audience, and therefore ad revenue. There was a big switch towards more neutral reporting and reporters signed up for 'journalistic integrity'.

Newspapers, often owned by big corporations, made heaps of money during this golden age with both ads and subscriptions, until a new technology came along that everyone wanted instead: Radio.

Newspapers were top dog in the 1800s.

They invented **mass market advertising** and made heaps of money.

Until radio came along and stole the crown.

Free news beamed into homes across the country through cheap crystal-set radios. It was live, with reporters on the scene sharing earth-shattering events like the crash of the Hindenburg real-time. Literacy was no longer a barrier to being informed. Companies like ATT leveraged their telegraph infrastructure into coast-to-coast networks. Presidents opted for radio over newspapers to reach voters.

But there was no way to charge for the airwaves. Anyone with a radio got the signal for free. Luckily, advertisers flocked away from newspapers and into the new medium, quickly discovering that radio ads worked... especially sponsored entertainment content like quiz shows, musical programs, dramas and mysteries. The Shadow, Ma Perkins, and The Green Hornet pimped cigarettes and detergent and breakfast cereal.

Advertising revenues for newspapers dropped nearly 40% in a decade, and lots of smaller papers folded altogether. The newspaper's golden age gave way to radio's golden age. Networks and stations gained tremendous influence on society and politics, and ad revenues rolled in for decades.

The influence of radio was so huge the Federal Communications Commission (FCC) enacted the "Fairness Doctrine," mandating that broadcasters had to dedicate some airtime to news for the public good, and that it had to be honest and balanced.

Then along came TV, and radio suffered.

Radio quickly became our favorite way to receive information. But unlike newspapers, the **only source of revenue was ads**. Shows had a cozy relationship with advertisers.

Then TV came along and we fell in love. The images and immediacy were just amazing.

By the 1950s nearly 80% of American households were gathering around the boob tube every evening. TV brought famous people, important stories, and presidential debates into our homes, and the Fairness Doctrine applied there too.

Like radio, they couldn't charge for airwaves, so television was paid for with ads. We wanted more and more programs, so the ad business grew with TV. Think Madison Avenue and Mad Men. Entertainment programs were specially created to pitch products and shown at the right time of day to find their market. Soap operas anyone?

News didn't make as much money as entertainment because it was dull. Worse, there was no way to pitch specific products at different audiences because everyone watched the same three reliable news shows. But the networks funded news anyway, because it was a public good.

In the 1980s the Fairness Doctrine went away, and TV learned how to make more money from news: turn it into emotional storytelling entertainment.

Cable news really benefited from the new freedom to embrace bias and opinion. They offered 24-hour talking-head news commentary like CNN and Fox. They could sell more effective ads with more precise targeting. The relationship between advertisers and media was tighter than ever.

Then the web came along and changed everything. Suddenly anyone could be a publisher or broadcaster.

TV was also funded entirely by ads. But the **"Fairness Doctrine" forced TV stations to maintain standards**, even if that meant lower ad revenue. That ended with cable TV.

The web arrived in 1994 and blew everything up. Suddenly anyone could be a publisher and could say anything.

The early web couldn't handle video, but newspapers quickly jumped in and put their articles online. They didn't charge for subscriptions because everyone said, "Information wants to be free." They assumed the advertising revenue would grow, so they sold those empty rectangles on their web pages to the ad-tech guys we talked about earlier. And they became even more dependent on sleazy ad-tech as their print sales collapsed.

The web eventually got fast enough for streaming, so TV joined the pack, jumping on the ad-tech train.

Then one day we all had a device in our pockets. It had a color screen and headphones and was with us 24x7. So we wanted news that would fit in with our mobile lives, constantly updated, entertaining, and nothing too long because... well, TL;DR. You know.

Today our phones deliver all kinds of news, in all kinds of flavors, with all different levels of respect for research and facts. Some give us long articles with evidence. Others give us shiny talking heads with loud opinions. Then there's the angry everyday Joes with their free websites, and the click farms writing flat out fantasies for the ad dollars.

Some have journalists and research staff to check facts. Others have studios and scripts and manicured spokesmodels. Some just pay high school kids to steal content and sell empty rectangles. Whatever the business plan, it all adds up. And who pays the bills?

Ad-tech and their sleazy trackers.

Once news started showing up on our phones, it was **game over for newspapers, radio & TV**. Or was it?

No matter how we get our news, someone needs to write it. And those writers, editors, talking heads and comedians need to get paid. **Guess who pays?**

Assuming we need news to make important, informed decisions, surely we don't just want loud personal opinions. Reliable, fact-based evidence would be more valuable.

Choosing good news is like choosing good appliances. We could just read the free reviews on Amazon or Best Buy. But those are just loud strangers' *opinions*, not real research. For all we know, the reviewer just liked the color.

Or we could go to someone like Consumer Reports. They rigorously test appliances, running them non-stop for months. They measure the reliability and provide clear comparisons of makes and models. It's expensive to do, and they don't accept ads. Would you even trust their reviews if they got paid by Whirlpool and LG? I wouldn't. That's why Consumer Reports ask us to pay—so there's no bias.

It's the same with news. Being well-informed is part of being a productive citizen in a democracy. News helps us make important decisions about important things: what state we might like to live in, which bonds to vote for, who should run the country.

So, we should be cautious of news that's just some celebrity talking head with opinions. And we shouldn't just accept the emotional rantings of individuals with no real expertise. Even calm, collected opinions from people who sound informed and smart are just opinions.

Because opinions are cheap to produce and easily exploited and manipulated to generate more clicks and money from ads. Evidence takes time and is not cheap.

If you want to buy a washing machine you can make your choice based on evidence, or opinions, or a TV talking head.

Opinions are cheap and funded by ads. Talking heads spew entertainment, which makes heaps of dough. **Experts gather evidence without bias**. It's expensive, so you'll have to pay.

Ad-tech tells news publishers to get our attention, and get us to click on more stuff, by spewing opinions that trigger our lizard brains. It's science, folks—when we're excited or titillated, our brains reward us with happy chemicals.

News that's designed to trigger divisive feelings gets more page views, more clicks, and more shares on social media. In fact, the more we share, the more ads get shown, the more data gets collected, and the more clicks get paid out to the publishers. Even if it's just an opinion. Even if it's fake.

For news sites funded by ad clicks, the cheaper their news is to produce in the first place, the more money they make off the ads. And what's cheap? Say it with me: Opinions.

Facts, on the other hand, are expensive. Sending journalists into hot zones to interview real authorities is expensive. Vetting and confirming sources takes time. Analysts and researchers cost real money.

There are still reliable news sources out there, doing that kind of real journalism. They might be dull, with no shouting and table pounding, but they present that expensive *evidence*, continuing the tradition of journalistic integrity and the spirit of the Fairness Doctrine. They don't depend on ads, so they're not skewing their news to capture clicks. Instead, they simply ask us to be fair and pay for the work they've done through subscriptions and paywalls.

Unbiased quality news is worth paying for. The guy who said "information wants to be free" also said "information wants to be expensive". Go ahead. Fact check it.

Emotional and **divisive news is the most profitable**, even if it's fake. We keep scrolling, clicking and sharing. Ad-tech loves to sell those ads over and over again.

Evidence is dull, with no emotions, no sharing, plus we have to pay. **Facts, not entertainment.**

Finding Quality News

To make the change necessary to enjoy news that won't clog your brain, start with the core of our thesis: **Free is bad.**

Pick just one classic news publisher who charges for evidence-based journalism that's not forged from opinion alone. One whose bias, if they have one, you like. Or at least can live with. It's likely a newspaper that's moved online.

Then install their app. Place that classic news icon on the home screen of your device. Open it when you're in the mood for news or interested in a subject and ready for some focused, unbiased reporting.

Go there to learn about important issues that only carefully considered, evidence-based investigation can illuminate. Vet the wackadoodle stories your emotional friends are sharing willy-nilly on social media.

But don't go there as a response to your own emotions, chasing the random stimuli that send you down unrelated rabbit holes.

Instead, turn off notifications, because you're going to give the news your attention on your schedule, not theirs.

And don't be discouraged or bored by the lack of inflamed arguments. That's what evidence looks like.

Finally, using ad blockers won't help. The problem isn't just the ads. It's the divisive news those ads fund. Block ads all you want; the junk news remains.

To escape fake news and garbage opinions, you need to pay. Find a publisher with a long history of reporting evidence and **subscribe**. Don't be fooled by free websites. Free Is Bad.

Install the publisher's app. Consume news on your terms: Turn off notifications. Avoid 'Breaking News' and clickbait. **Enjoy the sanity**.

Most likely the news app you installed has a paywall. I'll let you in on a secret: Life behind the paywall is rational, quiet, and populated by an informed citizenry.

The Reuters Institute for the Study of Journalism found the number one reason people subscribe to news is because it's "better quality than I can get from free sources".

Subscribers have left the world of ad-tech surveillance and divisive opinion. Paywalled news providers have no economic incentive to manipulate stories and sell rectangles, jerking our emotions this way and that to satisfy the needs of ad-tech, because we hand over a reasonable amount of money once a month in exchange for their hard work. No sponsored content, no editorials crafted to satisfy ad-tech, no fake news because there's no business need to have it.

We are the customer.

And by the way, I'm not here to change your politics. I'm happy with whatever news source you choose, with whatever inherent bias it has.

What matters to me is that you see the danger of news that's optimized to enrage and divide—just to sell ads. I hope we all agree that's a bad thing.

So it's my hope you'll make intentional choices about where you get your news, and you'll find a source that speaks to you, while also avoiding ad-tech and all its foul machinations and manipulations.

Because all I really need is to see my fellow citizens freed from the tyranny of free news.

TRUTH
THE PAYWALL
Reviews!
News
Ad
Ad

Summary

"*Information wants to be free*" said the web companies. But we can't actually have free stuff; someone needs to pay.

"No problem," said the ad-tech guys. "We'll pay. Just leave these empty rectangles on the pages."

They made more money by collecting more data, from emails and searches and our credit reports and everywhere. Ad-tech even made phones, getting even more data.

Publishers were hurting. "We can fix that" said the ad-tech guys. "Just write more polarizing articles. They're more popular. Look how much people share them!"

Smart people figured out that anyone can publish news. "Sweet," said the ad-tech guys. "More pages with more rectangles." This really worked well, driven by emotion and outrage. We shared it and liked it and those empty rectangles made them heaps of money.

Ad-tech was funding polarizing news because that makes more money. The Republic started to crack.

Some people started to wonder if news had to be like this. Was it always driven by emotion and outrage? Are evidence and opinion and entertainment really the same?

It turned out some news was based on evidence. We just couldn't see it because it was behind the paywall. Because it was funded by readers, not ad-tech.

Because "*information wants to be expensive*."

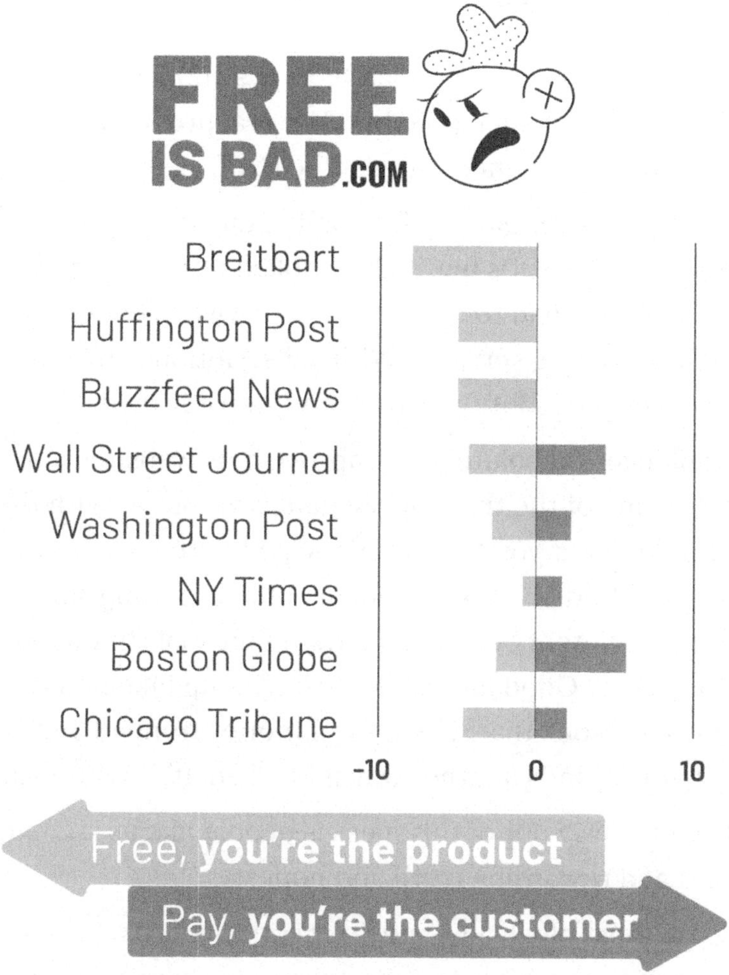

I analyzed 80+ news publishers and how much they depend on ads vs subscriptions. You can see if their motivation is emotions, opinions and outrage, or dull evidence. The complete charts can be found a **freeisbad.com/charts/** or in the unabridged book.

Using Ad Blockers

Plenty of people now use ad blockers and feel immune to the dangers this book highlights. The real problem is not ads, it's the way ad-tech incentivizes free stuff.

In 1999 Napster gave us free online music. We loved not having to buy CDs but it also choked off the money to musicians. They had to find a way to survive, which they did through streaming services. With subscriptions and ads that can't be blocked, Spotify thrived, and Napster died.

Ad blockers absolutely do improve how websites look and prevent some of the tracking shenanigans. But as with the music industry, if you take away the ability to make money, another method must be found. Ad-tech is getting smarter. Sponsored content? Can't block that. Editorials in the pocket of advertisers? Good luck there. Straight-up fake news? Nobody can block that. The flow of money from ad-tech is a river that just finds another way if blocked. It's unstoppable.

The only way out of this mess is people like us stepping forward and rewarding truth and honesty.

If you must block ads, and your conscience is clear, use Ad Block from eyeo which permits acceptable ads and therefore a trickle of money to publishers. Or subscribe to Scroll, giving ad free access to multiple sites.

For the important information about our democracy, please subscribe. Unless of course facts don't matter.

Ad blockers hide ads making sites look nicer.

But ad-tech fights back and gets smarter.

Think about it: **They make money somehow.**

You have The Power

Perhaps you got this book for free. I often run promotions and effectively give it away. I'm sure you've wondered why I would do this, when **Free Is Bad?**

I'm just trying to make the world a better place. For me, this project is *pro bono* or 'for the good'. Like Wikipedia, there's no hidden ad-tech shenanigans that's driving the agenda. It's just me, using what I've learned to help you.

If you want more history, evidence and insight, consider the unabridged version from Amazon. I especially recommend reading it if you find my explanation of News & The Media interesting or troubling. The unabridged version sets out my extensive analysis of 80+ news publishers and how much they are in bed with ad-tech. I also explain how I calculate the numbers. You can get a preview of this at my website **www.freeisbad.com/charts/**.

You can reach me on Twitter, Facebook and Instagram as @freeisbad.

John Marshall

I'm just trying to make the world a better place.

Made in the USA
Monee, IL
04 November 2026

67352884R00049